Big Machines for Fire and Rescue

by Brienna Rossiter

FOCUS READERS

SCOUT

www.focusreaders.com

Focus Readers is distributed by North Star Editions:
sales@northstareditions.com | 888-417-0195

Produced for Focus Readers by Red Line Editorial.

Photographs ©: Shutterstock Images, cover, 1, 4 (top), 4 (bottom), 7 (top), 7 (bottom), 9, 11 (top), 11 (bottom), 13 (top), 15, 16 (top left), 16 (bottom left), 16 (bottom right); iStockphoto, 13 (bottom), 16 (top right)

Library of Congress Cataloging-in-Publication Data
Names: Rossiter, Brienna, author.
Title: Big machines for fire and rescue / by Brienna Rossiter.
Description: Lake Elmo, MN : Focus Readers, [2021] | Series: Big machines | Includes index. | Audience: Grades K-1.
Identifiers: LCCN 2020033614 (print) | LCCN 2020033615 (ebook) | ISBN 9781644936696 (hardcover) | ISBN 9781644937051 (paperback) | ISBN 9781644937778 (ebook pdf) | ISBN 9781644937419 (hosted ebook)
Subjects: LCSH: Fire engines--Juvenile literature. | Ambulances--Juvenile literature.
Classification: LCC TH9372 .R67 2021 (print) | LCC TH9372 (ebook) | DDC 628.9/259--dc23
LC record available at https://lccn.loc.gov/2020033614
LC ebook record available at https://lccn.loc.gov/2020033615

Printed in the United States of America
Mankato, MN
012021

About the Author

Brienna Rossiter is a writer and editor who lives in Minnesota. She loves being outside and having campfires.

Table of Contents

Fire Trucks 5

Ambulances 8

In the Air 12

Glossary 16

Index 16

FIRE DEPARTMENT
RESCUE
HEAVY
RESCUE
IFD
INDIANAPOLIS
TACTICAL 14
fire truck

Fire Trucks

A house is on fire.

Fire trucks come to help.

They bring **firefighters**.

Fire trucks put out fires.

Their hoses spray water.

Their ladders reach up high.

ENGINE
hose

ladder
BRIDGEPORT FIRE DEPT.
E-12
ENGINE TWELVE
BRIDGEPORT FIRE DEPT.
BFD

Ambulances

A man is hurt.

An **ambulance** comes.

It takes him to the **hospital**.

AMBULANCE

The ambulance drives fast.

Its lights flash.

It has tools that help save lives.

AMBULANCE

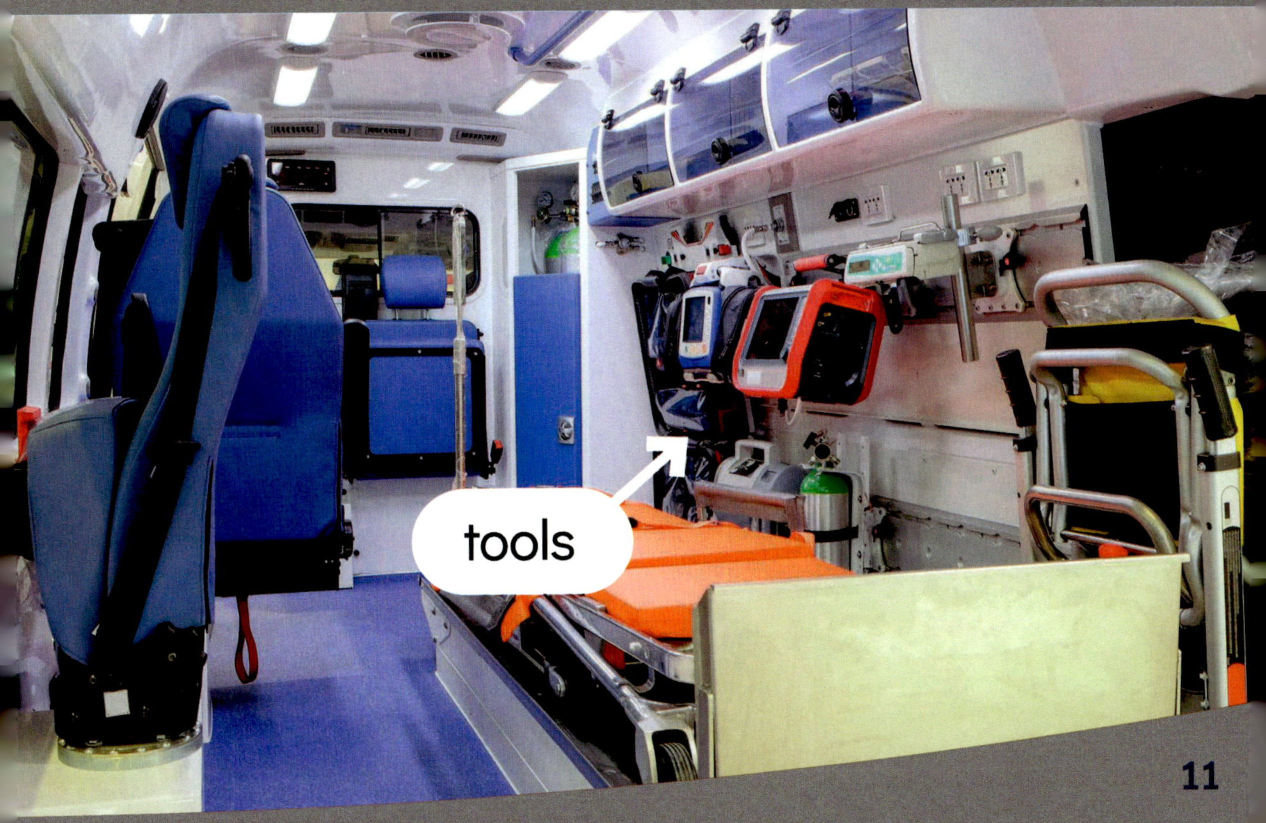
tools

In the Air

Some firefighters use airplanes.

Some firefighters

use **helicopters**.

They fly over the fire.

They spray the fire.

airplane

helicopter

A helicopter can rescue people.

It sends down a rope.

The rope pulls people up.

CZECH AIR FORCE
RESCUE
0715
rope

Glossary

ambulance

helicopters

firefighters

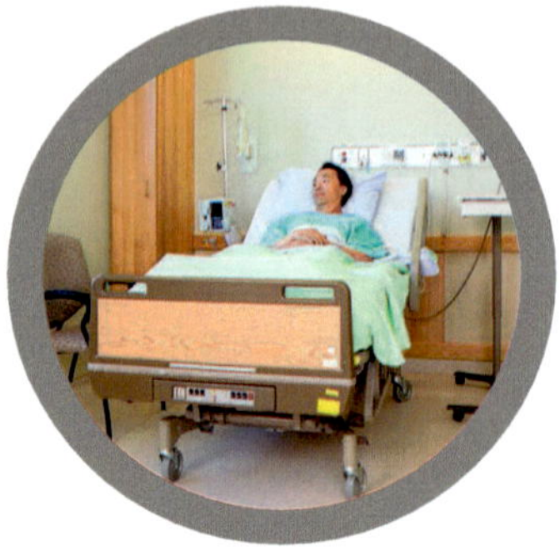

hospital

Index

A

airplanes, 12
ambulance, 8, 10

H

helicopters, 12, 14
hoses, 6